NASA and the Shuttle Shuffle

The disposition process for the Orbiters and how it all went wrong.

By
Jeannette Remak

©Phoenix Aviation Research 2012

You're a taxpayer, like me, right? Sure you are, and if you are an American who has lived through the United States Space Program or even if you came into it later on after man reached the moon, you know what the Space Shuttles are. If you didn't pay any attention at all to the space program, you still know that two shuttles were destroyed due to politics, bureaucratic fumbling and basic bad decision making. Fourteen people died in those two shuttles, so even if you didn't follow the space program, you heard about it on the news.

In the past two years, we have seen the demise and disposal of the U.S.'s only way to the International Space Station and the ability to orbit around our planet. The shuttles weren't made to go deep into space but they worked hard and worked well. However, since we paid for a huge portion of the International Space Station, including the Russian portion[1], we really do need to have a way to access it and retrieve our astronauts. As it is right now, we spend $60 million an astronaut to go back and forth via the 40 year old Russian Soyuz capsule in a deal made with Russia.

So, here you have a little background into what the shuttle did and now we will discuss how they were disposed of. As a taxpayer, I am appalled at just how this was done. There is a document on line called the *"Review of NASA's Selection of Display Locations for the Space Shuttle Orbiters."*[2] It is the NASA Inspector General Report on how the orbiters

---

[1] Thanks to Dan Goldin, former NASA administrator. When the Russians portion came up for payment, the Russians were broke. Goldin took funds from the Shuttle Safety upgrades and paid the Russians debt for the ISS.

[2] http://oig.nasa.gov/audits/reports/FY11/Review_NASAs_Selection_Display_Locations.pdf

were disposed of and sent to various museums. Now, some of us have had really hard and deep opinions on how the orbiters ended up where they did. You need to know that the process, like many things at NASA was seriously flawed and that is the reason for this monograph. I submitted a FOIA (Freedom of Information Act) to NASA for release of the documents supporting this debacle of the shuttle shuffle.(I will be sharing those documents at the end of this paper.) The reason that this shuttle disposition is so important to Museums, is that it has set a precedent for the release of government artifacts. The way it is right now, based on what NASA did, it seems like any agency can write its own ticket and do what it wants with taxpayer property and American history. The whole process reeks of different papers that support different legal languages concerning the ownership and disposition of real property that is by and large, taxpayer owned.

Essentially, the Shuttles belong to the American people. While NASA may have controlled and operated them, you the taxpayer, paid for that privilege.
Wouldn't it be fair to state then, that you, the taxpayer, should have a say as to where they should go in retirement. No, wrong. Once again, the American Taxpayer got screwed without even knowing it, except in this case " we wus' robbed!!" NASA, in their ultimate wisdom, set up the process of deciding which museum got what, based solely on their lack of expertise. Let's face it folks, NASA basically turned this whole thing into a competition for museums to own an orbiter, instead of an appraisal of the museums to host and support the artifacts. NASA did not even conduct a site visit to the various museums to see if they were truthful about the information that they

submitted. How can you justify this as a procedure equal to the task of protecting American history? That is the question posed here and that is what we need to find out.

Having worked in the Museum industry for some twenty five or so years, as a consultant, I have to tell you that this OIG report regarding the disposition of the shuttles is the biggest piece of fiction since the comic book was invented and folks, it is a sad comic book at that.

The process of disposition began this way: NASA had the shuttles and the program was closing down. Now, there are many people that disagreed with that scenario, and I for one, would have liked to see the program continue until a substitute was found. Constellation was on the books at that time in 2004, when President Bush made the announcement to end the program after the ISS was finished. However, when the Obama Administration came into being, Constellation was canceled and President Obama wanted the commercial sector to pick up the ball for supplying the ISS. There was nothing in the statement regarding the manned space flight program. Charles Bolden, the NASA Administrator, on watch during these proceedings, really is not the best administrator NASA has seen. While he is a shuttle veteran and all of that, he really needed to be replaced a while ago. Possibly the best of the recent administrators was Mike Griffin. Griffin was a soldier of the first degree, an engineer who worked the trenches and knew what was needed. Maybe he wasn't a top smooth talker, but Mr. Griffin had NASA at heart and he was good at what he did. He should have been around for the end of the shuttle program, instead of Mr. Bolden. Be that as it may, we have to deal with Mr. Bolden and his thought patterns or panders, I am not sure which just yet.

The start of the disposition process began with NASA writing an "ACT" that would ensure its final word would not be subject to any scrutiny. This was four years after President Bush gave the order for the program shut down. The Act called the " NASA Authorization Act 2008 -2010" was written to give NASA the authority to go ahead and place the shuttles where they felt it would do the most good, for whom, we're not sure of that either. In this ACT, NASA would be allowed, and allowed is a key word here, to get away with bypassing an older law called *"FPAS"*[3] "What this meant was that basically the *General Services Administration (GSA)*, yes the group that also spent a fortune on Las Vegas parties, would be the recipient of excess government merchandise and would distribute as warranted. It also meant that the government would seek a place for excess inventory within other government agencies, like the Department of Defense for one, *before* offering the inventory out for public consumption via agencies such as *AMARC*. As an example, *AMARC* is the *Aerospace Maintenance and Regeneration Center*. *AMARC* is part of the Air Force's Storage and Disposal Center. What happened due to this 2008 Authorization Act of NASA, it allowed NASA to make the decision for the shuttles, by and large, by themselves.

This Authorization Act also set the playing field for the *RFI*, or as NASA called it *"Request*

3.The program was created by Act of Congress in 1949 with the enactment of Public Law 94-519. This law enables excess Federal property to be donated to agencies which meet eligibility requirements.

for Information[4]" that allowed NASA to put out a request for museums to apply for an Orbiter. There is something wrong with this entire process and it should have been stopped at Congressional level long before it got as far as it did. One act like the *"2008 Authorization Act"* from NASA cannot supercede another act like *FPAS*. Why this was not mentioned early on can be due to a number of reasons:

1. A change in the White House
2. A change in Congress
3. No one thought enough about what was happening to do anything about it.

Much of what happened here goes back to the NASA mentality and how it got itself into disastrous situations due to its arrogance and compartmentalization practices. NASA believed it could do no wrong, regardless of the fact they had no expertise in museum theory. They had some of their best people, like lawyers and engineers and such, but *no one* that had museum experience and knew how to vet a host museum properly. We will follow through on that shortly.

As to the museums themselves involved in the Orbiter situation, there is no problem with the *National Air and Space Museum (Smithsonian Institution)*. It is the national depository for all things American and by rights should have an orbiter. They received the Orbiter Discovery. They also have the materials and manpower to maintain an exhibit the caliber of the Orbiter. So, we will say no more regarding the *National Air and Space Museum*, at least not right now.

---

[4] *The RFI form will appear in the appendix*

The National Air and Space Museum, Washington, D.C.

The Intrepid Bubble housing the Orbiter from their webcam on a rainy day in NYC. Note the "nipple effect" for the Orbiter's vertical stabilizer. See photo above.

Intrepid Sea Air and Space Museum   Let me open this section with a statement. Every other RFI I have read has its mission statement up front on the first page. Intrepid's mission

statement which goes as follows: " Honor our heroes, educate the public and inspire youth, appeared after its marketing and promotions statement.

Intrepid certainly did not belong in the first cut of museums in NASA's first RFI. Why? According to the NASA report, the *Recommendation Team* concluded, without a site inspection, that Intrepid could handle the Enterprise. How NASA couldn't figure out that Intrepid was an aircraft carrier to begin with really blows the mind! How did they figure they would get it on the deck? Via the stairs? The shuttle now sits on the aft part of the flight deck(over a WWII bomb elevator), in a "environmentally controlled tent" seventeen stories above the Hudson River. Have you ever seen a Nor'easter[5] in NYC? The winds that occur during a storm of that nature have reeked damage on the flight deck. Having restored an aircraft on that deck for six years in service to the USAF Museum, I can tell you it's dangerous. I have been on that deck during snow and ice storms, hurricanes, nor'easters etc. It is a dangerous place to put a tent, especially one holding a precious artifact. Intrepid had plans for purchasing a parking lot on the other side of 12th Avenue. However, we haven't heard much from the TLC (Taxi Limousine Commission and NYC) on that part, but it sure sounded good when they pitched it. It is hard to justify how NASA could allow an Orbiter, even a atmospheric orbiter, into a situation like this. There was no place to put the Orbiter initially and there won't be anywhere else in the near future. We have also heard that they planned some other stunts. You must know that Intrepid is used like an entertainment center instead of a WWII

---

[5] Please see photograph at the end of the Intrepid section. It says it all.

museum with an enormous battle history. Name the day, Halloween, Bastille day, Intrepid has a special event. Now, people may say that things have changed. Intrepid has been the butt of jokes from other museums which called it the "Carnival Cruise Lines of Museums. They have hosted events like the Robby Kneivel[6] jump over vintage aircraft, boxing matches on the flight deck in which a boxer passed away because EMS couldn't evacuate him quick enough due to the elevator to get down to the pier[7]. The gentleman lingered in a coma and finally passed away. There have been many side shows like this on that hallowed flight deck where WWII Naval men and Marines were killed in a kamikaze attack. It has been a sore point on how Intrepid has been handled.

Yet, to be fair, after a particular incident regarding the A-12 Blackbird, and the owner of the A-12 Blackbird, the National Museum of the U.S. Air Force, Intrepid was told to straighten up its act. A concerted effort was made but not enough for my taste and other historians who know the story. However, let's face it, NYC wanted a "meal ticket" much like the A-12 Blackbird, that now sits on the side of the flight deck, with a wing hanging over the fence, once was. If it's sexy, Intrepid wants it to bring business to the Westside amusement area that NYC has been working to build up. One of the reasons that Intrepid was moved out a couple of years ago and finally got a paint job was because NYC didn't want an eyesore on its new park land.

As to Intrepid history in the Space program, she was primary recovery ship for the first

---

[6] Photo of jump shown in appendix

[7] appendix will carry stories and images of Intrepid's many "events"

crewed Gemini 3 mission landing and Scott Carpenter's Mercury mission and that's all folks. This was hardly an astounding history for the Intrepid/ NASA space program connection nor was it enough to hand an orbiter over to them. How you can equate this history with Houston's years of mission control history, is beyond the realm of reason. As much as I do love the history and glory of Intrepid and the men that served with her, her masters, just like zebras, never change their stripes. They were and are totally unworthy of Enterprise and my only wish is that nothing serious happens to the orbiter. Enterprise already took a hit while being delivered by the barge, which ran into a piling and damaged a wing.

Photo (NASA/Intrepid) How the shuttle is housed on the Intrepid. Note the vertical stabilizer right at the roof of the tent. This tent is 17 stories above the Hudson River on the ship's aft flight deck.

This is an artist's concept of the final shuttle Enterprise home on the deck of Intrepid. What ever happened to all the other plans that were put on the RFI, then dumped after the announcement of the award of Enterprise?

A photo of the Enterprise after the Hurricane Sandy hit in Oct 2012.

In closing this section on Intrepid, only photos can speak louder than words. During the massive storm that hit New York City on October 29, 2012, the " bubble tent" lost electrical power and the Orbiter was subjected to 90 mph winds and water and wind damage. So much for safety and thought in placing historical artifacts, this is a travesty that should not have happened.

The California Science Center is an enormous complex housing many exhibits including an Orbiter which is still waiting to be displayed. (Photo California Science Center)

Another view of the California Science Center (Photo: California Science Center Website)

California Science Center is next on the list. The CSC is a nice enough place, but it doesn't know anything, nor does it have any provisions for caring for a flown orbiter, the rarest of birds, at least one that doesn't have feathers. The center is a huge complex that houses an eco system, historical events like a Cleopatra exhibit, an aquarium and various other exhibits. The reason CSC received and orbiter just shows what went on here. The CSC is twelve miles from where the shuttle was built. Again, hardly a satisfying reason to give them an orbiter. Right now, they are begging for funds to build a facility to house the orbiter that is living in a tent (sound familiar?). This is a much more serious situation, however. Endeavor is the youngest shuttle and it has worked in space. This is a piece of our space history that can not be recovered should anything happen to her due to lack of maintenance and care. Endeavor is the last shuttle to be delivered somewhere around October 13th, 2012. After landing at Los Angeles International Airport, she will be towed through the streets of Los Angeles on a twelve miles tour of the city, chopping down a boulevard of ten year old trees, along the roadway as she passes, due to width issues, before ending up at the California Science Center. (*Good Planning Environmentally Conscious California!!)* It's just another road show, another circus event. This is just the type of "party" that needs to be thrown for the last shuttle built. No one is sure when this shuttle will be put on exhibition.

From the NASA website: how KSC plans to display its orbiter in a new pavilion under construction.

Kennedy Space Center is the next on the list. Atlantis was given to Kennedy Space Center. Hopefully. NASA will have enough funding to finish the pavilion its building to house the orbiter. I can't complain about this placement. This was the only one that made any sense, next to the Air and Space museum. KSC is the place that shuttles called home. They were launched from and landed here. At least, NASA did this right.

To add to this, in the OIG Report on the Shuttle Disposition, it is obvious in the table matrix developed and scored by the NASA Recommendation Team[8], shows just how inept and isolated this team really was. According to administrator Bolden, that was just the way he wanted it. Really? First, the information received in the first RFI was questionable.

The Recommendation Team did nothing to check and see if it was accurate. Even museums make mistakes. The Charles Bolden idea of putting the shuttles in a place where

---

[8] All tables will be reproduced at the end of the paper in the appendices

they could be seen by a vast population of international and local populace, is not only unrealistic, it is questionable. The bigger question is how did Bolden get into the position of being the final decision maker? Bolden claimed he wanted education to be the theme, with math and design a close runner up. Bolden's idea that geographic locale and international population were first and foremost in his mind, really should have been questioned. He never thought of how the shuttles would be maintained and supported after these venues received the orbiters. It takes money to do that, something most museums don't have. To be quite honest, I am not sure he cared. Better that visitors had a chance to see what NASA and the U.S. was no longer flying and show international visitors that the U.S. was now out of the manned space program. Better yet, that 5 years down the road when the Orbiters need maintenance and upkeep, there would be no way to service them.

Bolden also made damn sure that none of this information would leak out as far as who was going to be awarded what. The NASA budget was in Congress and the last thing he wanted to do was tick off a Senator or two. Bolden held back the award notice twice. It seemed to me that all this disposing of the orbiters went down pretty fast, with no turning back. Why? Well, I am not sure what the answer to that is, other than perhaps some political ploy.

There are flaws in Bolden's thinking regarding how museums operate. The museums that charge admission, really don't know how many international visitors that they get. In reality, no museum free or admission, unless it has people standing at the door and asking how many people are from out of country, will ever know. There is always a guesstimate, but no true figure, not unless someone is physically counting. That leaves us with this thought, how could Mr. Bolden use this idea as a pre-determining factor for who received an orbiter? It's not valid means of approval for receiving a precious artifact. It looks as if Mr. Bolden lumped in geographic location and the *hope of*-international visitation, but he missed the much more serious point of whether the museum had the means and wherewithal to support the orbiter pass the first year on exhibition. Time takes a toll on all static displays and it won't be long before these artifacts will need attention. Geographic location and populace is *not* a defining issue when placing national treasures.

The Winners

Let's take a minute to look at the locations of the Museums themselves:

Intrepid- New York City: Intrepid for those not familiar with NYC, is on the west side of the New York island, 12th Avenue and 48th street. She sits in the Hudson River. First, Intrepid is charging a separate fee **IF** you *don't* want to see the orbiter and that is $26 for Adults, $20 for kids 7-17 etc. Intrepid *does* charge a higher fee of $30 for the Shuttle pavilion and ship --adults entry, kids are $23up to 7 thru 17, little kids $16,Veterans are $23. $30 is a lot of money, especially if you have children. To be honest, who is ready to tell their child they **won't** be seeing the space shuttle. Intrepid can't be making out too well in this economy. Basically, you the taxpayer, are paying twice for the privilege of seeing the orbiter. Intrepid paid $8.3 million for the Orbiter and that was most likely raised privately. On the matrix table submitted by Intrepid to the RFI team, they claimed $22 for entry fee.

Intrepid brought the shuttle to its pier via its "famous" barge program. Any aircraft that Intrepid gets, usually comes in on a barge. In this case, since there was no onsite investigation of the facility until *after* the award, by NASA's Recommendation Team, the shuttle was damaged on its incoming float down the Hudson River, by banging into a piling due to high winds and damaged its right wing. Intrepid received a 10 on the RFI matrix which means that transportation risk was low.

Another issue worthy of bringing up at this time has to do with Museum accreditation. On the initial RFI, Intrepid was listed as being accredited by the American Alliance for Museums and the Smithsonian Institution. This is the gold star of the museum society and a highly coveted one at that. It is very difficult to receive accreditation. In fact, it was later noted by the OIG report, after the second RFI came in with the same information on the matrix, that Intrepid was in fact not an accredited member. The NASA Recommendation Team was not checking into the answers it was given in those RFIs.

In checking those RFIs, the FOIAs produced some very interesting facts
Intrepid's RFI was a total of fifteen pages as opposed to the other twelve RFIs that ranged in anything from twenty five to ninety two pages with explicit information on how they would do things if they got a shuttle. Something else needs to be added here regarding that RFI. It was written in 2009 at which time the CEO of Intrepid was Mr. Bill

White. Mr. White had been on the Intrepid executive staff for some two decades. Mr. White resigned his position as CEO in 2010 after an incident with Attorney General, now Governor of New York, Andrew Cuomo. According to a Wall Street Journal Article dated May 20, 2010:"*After he was found to have been operating as an unlicensed placement agent, a position marketing financial investments to pension funds. Bill White & Associates allegedly received a $2 million fee from City Investment Fund L.P., a real estate investment fund co-sponsored by real estate company Fisher Brothers that received pension fund business. A spokesman for the attorney general's office confirmed that Mr. White had been subpoenaed, but declined to comment further on an ongoing investigation*". White resigned from Intrepid and according to what the Intrepid Board said, he would not be permitted to speak about his time on Intrepid. This was based on a separation agreement. White also paid back $2 million to the Attorney General to keep himself out of prison . While NASA said that all of this had no bearing on if the Intrepid would receive an Orbiter, looking at the fifteen page RFI that was submitted, it would seem not. The other RFIs were loaded with facts and figures , tables and demonstrations of who and how they would manage an Orbiter, but Intrepid's showed none of that. The RFI is attached in the appendix with all the other museums that applied.

As to the educational part of the RFI, Intrepid was proud to cite that many activities that had nothing to do with education and the public access to it. The *SANYA Initiative* ,which was intended to create increased communication between senior ranking generals of the Chinese and American armies was a highlight of their RFI. It also boasted the American/

Israeli Chamber of commerce and the NBC 4th of July fireworks celebration. The public was able to access that so they could sit on the flight deck and watch the fireworks on 12th Ave on the Hudson which was covered by TV. Speaking of TV, how about the NBC TODAY show, the NBC Football Night in America, CBS Early show, etc. ad nauseum. It's quite strange that I don't see other NYC museums of prominence like MOMA and the Metropolitan Museum of Art, with the exception of a red carpet charity function, getting that kind of coverage. Intrepid's RFI actually had a Marketing and Promotion section which did not occur in any of the twelve RFIs submitted.

The requirements for exhibiting a shuttle were hardly handled in the detail that so many of the other Museums took time to work out. The hopes for the aforementioned parking lot deal was not to be and we are left with that abortion of a bubble tent on the flight deck.

Their RFI claimed Intrepid's expertise with moving large aircraft like the SST. Well, the SST was damaged and British Airways was not too pleased about it and as already mentioned, the Orbiter Enterprise had its right wing slammed into a piling when it arrived at the pier. So much for the expertise. If you look at the numbers of how many visitors have gone to Intrepid, they are quick to point out that 45 million people are in and out of NYC all day long. It doesn't quite answer the question.

There is one question that is left, however, how did Intrepid get an Orbiter and Houston/Johnson Space Center, the National Museum of the US Air Force, the Adler Planetarium, Evergreen Museum and various others that will be addressed in the Losers section are still asking why they didn't? These other museums had the money, the wherewithal, the expertise, yet, they were left out in the cold without so much as an explanation.

While I have no reason to question the Intrepid Museum, the ship herself and her crews are marvelous and heroic in serving this country, that is never to be doubted or debased. I do question the rather obvious and blatant use of marketing tricks and promotionalism used in their RFI. I don't see the historic trust that should be promoted. I see the next "Big Ticket" item that will be used to draw people to the museum. That's not what its all about, folks It's about the history, the preservation of that history, the promotion of education and pride, not a reason to have a "special event".

California Science Center: is the next on the list. The CSC is at least honest enough not to charge an admission fee for the museum. It is trying to build a new facility for the orbiter and is taking donations. There is not much wrong with that, except that the CSC really hasn't the facility in place to maintain something like a flown orbiter. This artifact needs maintenance and care and this cannot be stressed enough. It's not just enough to house the artifact, the care of the artifact in coming years, when there are no parts and no one who knows and understands these creatures brings us to the real crux of the problem that NASA created.

National Air and Space Museum: There is no reason to really mention NASM. You don't get much better than the Paul Garber Center for restoration and the Udvar-Hazy Center for display. Again, NASM does not charge admission.

The Kennedy Space Center: This is the true home of the shuttle. KSC is now in process of expanding, to create a pavilion for the Atlantis Orbiter. This bird really lucked out. While KSC does charge $45 for entrance fee, it does include an air-conditioned bus tour of the large facility and much more. If KSC can't maintain its orbiter, things are really bad.

KSC's purpose is to "tell the NASA story and inspire people to support the exploration in space". KSC is the " nation's gateway into space."

KSC educational mission was developed in both formal and informal programs, reaching both school and youth groups. The programs serve some 90,000 schools and youth groups and 100,000 adult guests. STEM (science, technology, education and math) is also a major factor here.
According to KSC's RFI, they have an amazing array of educational programs:
Camp Kennedy Space Center (1 week group)
a. Overnight adventure (school and scouts)
b. Salute to home schoolers and scouts
c. Brevard (county) Space Week (6th grade)
d. Brevard Science and Technology (7th grade)
e. Student Astronauts(1 day training)
f. Astronaut encounter (spend day with astronauts in briefings)
g. Educator Study Passes (annual free admission for teachers)

KSC' s RFI is 40 pages of information on how they will build the new housing for the shuttle and support the shuttle after she is installed with educational programs and such.

The Losers

We need to take a look at the museums that were overlooked and why they were overlooked. In the OIG Report, it was noted that the National Museum of the U.S. Air Force was dumped from the running. The reasoning behind this, according to Mr. Bolden, is that they didn't have the money up front. Neither did the other museums, including KSC. In fact, the USAF Museum had a $14 million appropriation given to them from Congress, for the direct purpose of getting a shuttle. They would have had what is known as an interservice transfer that would have allowed for the rest of the $28 million. Mr. Bolden, not understanding interservice transfers, or not wanting to, immediately dumped the USAF Museum for lack of funds. According to the first and second RFI submitted by NASA, the information that the Recommendation Team added scores on to was incorrect. The Recommendation Team said the USAF museum couldn't meet the delivery schedule but since it is the USAF, I find that hard to believe. They could have flown the orbiter right into Wright Patterson AFB and towed her over to the museum.

Next, they said the USAF Museum was not affiliated with the American Alliance for Museums, in fact, they were. They also placed the USAF Museum in the west for geographic location instead of mid-west. And again, there is the little item of a federal law that was completely ignored by the Administrator of NASA. As to fees charged for entering the museum, there are none. It is free of charge museum. The museum has ample space to house an Orbiter easily, and the restoration department is one of the best in the world. It also has free and ample parking. The Museum is a few miles from Dayton International Airport, housing many airlines and it does see its share of international visitors, most likely no more or less than any other museum with exception of the National Air and Space Museum. Of course, all this was rectified in the OIG report, which was after the announcement of award was made. I need to add here, that there are many astronauts that were USAF born and bred. They died in both tragic disasters of Challenger and Columbia. It seems to me that the NASA Recommendation Team was so isolated that they couldn't see their way into a sunspot, if they had too. Blind is another word that comes to mind.

As to the parties that formed this "Recommendation Team", the list goes something like this:
Sue Kinney- director of Logistics Office of Strategic Infrastructure
Robert Sherouse-Transition Mgr, Office of Strategic Infrastructure
Jonathan Krezel-Space Shuttle Office Transition and Retirement lead.
Mark Batkin- Attorney Advisor, Office of General Counsel
Courtney Graham-Associate General Counsel, Office of General Counsel
Mike Curie-Public Affairs specialist, Office of communications
Rick Itrving,-Legislative Affairs specialist, Office of Legislative and Inter-governmental Affairs.

I am sure you see no one with any museum experience that would have qualified as a valid officer to help this team. As usual, it was for NASA, by NASA and tunnel vision set in once again. Mr. Bolden's concept for the Orbiters was lame. There were many good, solid museums, that would have been more than worthy to house an Orbiter. Yet, Mr. Bolden superceded that possibility. At the Congressional hearing, Senator Sherrod Brown of Ohio contested Bolden's concept and possible choices, Bolden was arrogant and already had firmly made up his mind. All he had to do was sign the paper, which he told Sherrod Brown he would do right after the hearing.

The argument of whether the Orbiters could have done better for retirement homes is without question a valid one. I firmly believe that something should be done regarding this process that NASA created to serve itself and its interests. While it may not change things for the Orbiters, since the SCA[9] and the rest of the equipment needed to move them is gone for good, the process that created this mess must be brought into the light. We can hope for changes of venue, but seriously doubt it will happen.

Inside the huge Seattle Museum of Flight in Washington state

---

[9] SCA Shuttle Carrier Aircraft along with mating and de-mating equipment

The Seattle Museum of Flight This museum is a consummate aviation/aerospace museum with a long history of working with the Boeing Corporation. The museum was started in 1965. They have one of the most beautiful displays of the MD-21 Blackbird and they worked like dogs to create it. In fact, they work like dogs at everything they do. Much like National Museum of the U.S. Air Force, the Museum of Flight has more than ample space for exhibition purposes. It also has ample support by Boeing and the Washington area. It also has one of the largest aviation/space libraries anywhere. The curation staff and the restoration staff, like the NMUSAF, is magnificent with 132 paid and volunteer staff. How come this museum was dropped from the running in getting a shuttle? According to their RFI, they had all the requirements: climate controlled exhibition space, they were ready to receive an orbiter just as NASA wanted within the time frame, and they had the money. Their RFI was the one of the best that I read. They are also a member of the Alliance of American Museums and maintain the standards necessary to hold that title.

There is no understanding as to why a museum like this was dumped on the side to allow a place like the Intrepid to receive the Enterprise, which by the way the Museum of Flight would have been happy to take. They have a runway adjacent to the facility to receive the shuttle. Below is the list of items that MOF had to offer historically via their contact with Boeing[10]:

---

[10] The RFI for MOF will be in the appendix.

a. The Boeing Company was founded in Seattle nearly 100 years ago and is not the dominant commercial aviation company in the U.S. It is the primary contractor for the International Space Station and part of the United Space Alliance (USA) which supports the ISS and the Shuttle.
b. Boeing built the first stage of the Saturn V rocket which aided in getting to the moon.
c. Boeing Seattle built the Lunar Rover which remained on the moon to this day.
d. Boeing built the first robotic lunar orbiter which photographed landing sites for Apollo.
e. Boeing built the 747 which transports the Space shuttle : the SCA Shuttle Carrier aircraft
f. The President of Rockwell International, which built the Shuttle, Mr. George Jeffers, was a graduate of University of Washington College of Engineering.
g. The Space Shuttle aerothermal characteristics were studied and created by Professors, John Bollard and James Mueller of the University of Washington, College of Engineering. Their work contributed to the successful development of the ceramic tile thermal protection system
h. Boeing Seattle built the Inertial Upper Stage (IUS) which was carried in the Space Shuttle payload bay 15 times and launched large satellites like the Chandra X-Ray observatory.
i. There were 25 astronauts that hailed from Seattle including STS-51L commander Dick Scobee and Col, Mike Anderson from STS 107.

I could go on but I feel the point has been made. Boeing supports the Museum of Flight without question. When you lay this list out against the Intrepid's ONE Gemini pick-up, there is hardly any comparison at all. MOF should have at least received the *consideration* of the Enterprise if not a flown shuttle, which it truly deserved,

March Airfield Museum in California

March Airfield Museum:
Next on the Losers List, we have the March Airfield Museum in California. There is another question before we get into the whys and wherefores of March's dumping. When we look at the March Airfield location and the California Science Center location, there comes the issue of about 400 trees that were cut down to allow the shuttle to pass on the roadway to CSC. This has to be the most stupid, ridiculous way of bringing a shuttle into a museum. 400 trees had to die to allow the shuttle to pass through? Does this make sense, when there is a museum with a better RFI available to the California area? The only reason given for why CSC got a shuttle was that is was twelve miles from where the shuttle was built? It's just another reason why this whole process needs to be made transparent.

Back to March Airfield. March[11] also had a very good RFI and met all of the requirements., right up to being able to house

---

[11] The RFI for March Airfield will appear in the appendix

the Orbiter indoors by moving six of their aircraft to another education center until the "Columbia Memorial" could be built. Again, Intrepid has the Enterprise in a Bubble tent, on a flight deck 17 stories up in the air..

This museum had the means, money and wherewithal to support a flown orbiter, yet would have been happy to have the Enterprise. They too had the means to support an Orbiter after the blush of the first year had worn off and better yet, California would have gotten a shuttle and 400 trees would have been saved.

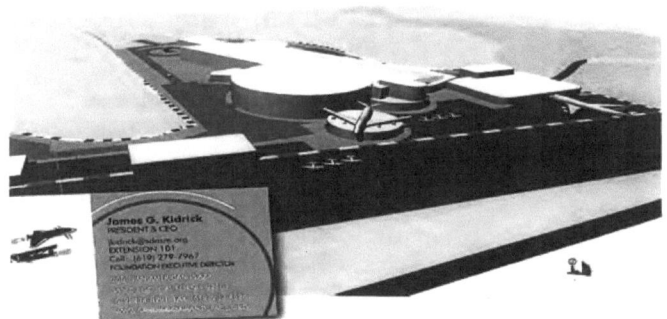

Plan for shuttle housing at Brown Air field

The San Diego Museum of Flight

San Diego Museum of Flight While the San Diego Museum of Flight is not as large as the NASM, MOF or the NMUSAF, it does have a very credible background and there is no reason why this museum should have been overlooked for a Shuttle. The museum is one of the oldest, beginning in 1963, is an accredited member of the Alliance for American Museums and upholds those standards. They had a very impressive financial background and support is solid. With Brown Field right next door, they had a safe landing spot for the shuttle and the SCA, without having to move far to the indoor exhibit area. They also boast a restoration crew and a solid advisory board. Why would they not be considered? [12]

---

[12] The San Diego Rfi is in the Appendix

Image of how the NMUSAF would accommodate a shuttle and the entrance to
The NMUSF

The National Museum of the United States Air Force

While we have spoken somewhat of the NMUSAF. There is something that should be shared with you, here and now. This is the letter that was included in the RFI that was FOIAed out of NASA. It reads:

"The United States Air Force, under the provisions of 40 U.S.C. 524(b)(2) requests the interagency transfer of a shuttle Orbiter, with SSME which may be excess to NASA mission needs to fulfill a requirement for the National Museum of the

United States Air Force (NMUSAF). The Orbiter is required in order to fully develop the growing space collection of the museum and present to the visiting public the USAF/NASA partnership and the contributions of the Shuttle Orbiter program to National defense. The Air Force past and current contributions to the shuttle program are significant and touch all aspects including research, design, engineering, astronauts, and mission focus. The Orbiter would serve as the centerpiece of the space collection along with a Titan IV space launch vehicle, Mercury, Gemini, and Apollo capsules, and other space artifacts and exhibits for public enjoyment and education.

While the USAF would be pleased to receive any available operational shuttle, there is a strong preference for the shuttle Atlantis given its close association with specialized Department Of Defense (DOD) missions and personnel. The NMUSAF is centrally located in the American heartland on Wright-Patterson Air Force Base, in Dayton Ohio."

The working sentence here is : "The United States Air Force, under the provisions of 40 U.S.C. 524(b)(2) requests the interagency transfer of a shuttle Orbiter, with SSME which may be excess to NASA mission needs to fulfill a requirement for the National Museum of the United States Air Force (NMUSAF)." Interagency transfers are part of the GSA/ USAF/ NASA law. Federal agencies do this all the time. It's not a new practice. Considering the amount of blood that was shed in the Shuttle Program, much of it was USAF Airmen and Airwomen who shed their blood, it would have been only fitting to give the Orbiter to NMUSAF to honor those folks. The NMUSAF has monumental resources and their staff is one of the best in the world of museums. If that wasn't a big enough answer, then how about this, it's the law ! Yet, it is a law that Administrator Bolden chose to ignore outright. That is exactly what he did when he decided to dole out shuttles to museums that either did not deserve the honor and can't support the shuttle. Three shuttles are on the east coast within a couple of hundred miles of each other. There is no reasoning to that, unless you want to call it politics and let's face it, that is part of the problem.

The Adler Planetarium in Chicago

Adler Planetarium

The Adler Planetarium has been around since 1930. Their idea for a shuttle would have been to build a facility for Chicago's Space Science Center " to inspire the next generation of explorers and become the world's premier space science center." They felt that an Orbiter would spark visitor's enthusiasm for space exploration as well as encourage youngsters to pursue careers in science, engineering and technology.

Adler wanted to work with the City of Chicago, and the state of Illinois to raise the necessary funding for the project. Adler also wanted to construct a building adjacent to the museum in the city of Chicago parking lot. Now while that does sound familiar, ala Intrepid, Adler has a great history demonstrating successful expansions of this nature.

Adler has an enormous outreach in the educational department ad there are may after school and Saturday programs for kids of all ages.

Now while Adler may not be on the tip of your tongue for "hot" museums, lie the Intrepid or CSC, due to their large public relations and advertising departments. Intrepid boasts a huge one. Adler would have been a prime site for a shuttle. Why? It had everything, money, expertise and good location. Smell something funny yet? Yes, I thought so. I've been smelling it since the awards were made.

Bush Library and Texas A&M University Campus

Bush Library and Texas A&M

Bush Library and the Texas A&M University teamed up to make this presentation for an Orbiter. The museum, in Brazos Valley and part of the A&M University would later expand to include a center for a Space Policy Institute. They had a plan for donations through the public funding grants and limited use of public money. Many alumni of the A&M University also were eager to help financially.

The Orbiter would be displayed in a guarded building specifically made for preservation and display. The Orbiter would be visible from the street level and placed nose forward towards the entrance. It would be curated by the George Bush Library and Museum staff, who are already dealing with many significant and rare items for the Library.

It would go without saying what a boon it would be for the Orbiter to be on view on the Texas A&M campus. Education could not be ore prevalent in the range of plans. The main objective would be to build the Space Policy Institute and Learning Center around the Orbiter display. The mission is to inspire exploration into the unknown frontier of space.

The SCA (shuttle carrier aircraft) could land 1/4 mile away at the Easterwood Airport. The Orbiter would be used to bring both the educational and scientific community together. This would also create jobs in the area and bring with it, millions of dollars of revenue to the local economy. According to their plan, the money would have been delivered on time and they were ready to receive the Orbiter at the time set by NASA.

The good thing about the Brazos Valley location, which is where the Bush Library and Texas A&M reside, is that the climate is friendly. It doesn't have the problems that Houston, which is sometimes subject to tropical storms, floods, humidity etc. has. The corrosive effects of wind blown, sea water air is not found in the Valley.

Texas with approximately twenty four million people is the second most populated state behind California and beats NYC by approximately five million people. It is within a five hour drive of many of the southern and western states.

Texas A&M University has a long, strong association with NASA and the shuttle program. The RFI goes on to state that its policies on loaned artifacts policy, exhibition policy, exhibition procedures, research policies, accreditation with many associations including AAM. It goes on to state its educational policies, financial stability, attendance figures, it's estimated income and a very reasonable general admission price. In its thirty five page RFI, you get a clear and concise vision of how this group would have handled an Orbiter.

The Evergreen Aviation Museum in Oregon, houses the Howard Hughes Spruce Goose.

Evergreen Aviation and Space Museum This museum's mission is on the first page of its RFI. Their statement, " to inspire and educate, to promote and preserve aviation and space history", gives its total perspective. They convey that statement by the world class display of aircraft and artifacts in their care. EASM's educational programs are as world class as their exhibits. Their most spectacular achievement is the movement of the Howard Hughes H-4 Flying Boat, known as the "Spruce Goose". Talk about moving large aircraft! They did a fantastic job. Their collection starts with the earliest of aviation and goes on to the SR-71 and into space with the Titan II and IV missiles, including Mercury spacecraft. Evergreen has a history of working with NASM, NMUSAF, NASA and the Kansas Cosmophere to acquire artifacts.

EASM has a solid commitment of caring for their collection. They have proven their expertise in moving and displaying large aircraft. The Orbiter would have proven to be no problem for them and possibly the lives of 400 trees in California might have been saved if this museum received the Orbiter instead of CSC.

Evergreen's Space Museum Building is a centerpiece in educating the public in man's space efforts. EASM believed that bringing the Orbiter to their institution would be in keeping with their statement," To inspire, educate, promote and present aviation and space history and honor the patriotic service of our veterans". Evergreen's programs that are designed for all age levels is actively promoting education in science, technology, education and math (STEM) with the goal of being the " M.I.T. of the West" for aviation and space science. It already has a faculty of 17 accredited educators and houses a STEM focused high school. This school works with colleges and universities nation wide, to promote all these disciplines which is critical if the United States is to remain as a world leader in technology.

Fundraising for the Orbiter project would come from the Evergreen family of companies. (EA) Evergreen International Aviation made a solid link between NASA and Oregon.

The NASA Shuttle Carrier Aircraft(SCA), a Boeing 747, has been involved with Evergreen in the maintenance capacity since 1991. With the solid funding behind Evergreen Museum, it can plan and accomplish its goals, a position that many non-profit museums would love to be in.

EA was prepared to give the NASM, NASA or the Department of Defense, a substantial contribution of a cargo airlift or maintenance. They believe that it is important to give back to the government and offset or trade of services would have been beneficial to all.

Others pledged support to EASM to bring an Orbiter home. The Oregon Museum of Science and Industry and Hiller Aviation Museum were happy to help.

EASM is a debt free organization with property, structures and artifacts in excess of $200 million. EASM opened its new Space Museum on June 6, 2008. It is climate controlled, and a state of the art exhibition space. EASM would have had no issues in meeting NASA's schedule for funds or delivery. Having moved the Spruce Goose, the shuttle would have been no issue at all.

EASM would have been a wonderful place for the Orbiter to represent the West Coast. The RFI goes on to cite the many educational programs available to the public:
a. Evergreen Youth School: for 16-21 yr old students. It is a ground school for future pilots.
b. Scouting: By and Girl Scouts would gain merit badges for aviation, orienteering, astronomy.
c. Civil Air Patrol: cadets are offered classes and gain a chance to win merit badges.
d. Discovery Ambassadors: 12-18 year old students provide classes for communications and presentation skills through hands on demonstrations for museum visitors.
e. Aerospace Book Club: for K-3 yr olds to excite and challenge the very young in aviation and space.

The plan for the Orbiter display would allow the public to walk around the Orbiter. With specially trained docents available to demonstrate how the Orbiter was used in space via graphics, photos, artifacts models, etc.

The RFI continues to explore the museum practices regarding loans, refusal of offer, accessioning and other viable museum procedures.

Basically, the museum would have been a excellent choice as a west coast home for the Shuttle. EASM literally had it all, money, expertise, motivation. So, why weren't they considered?

THE HOUSTON JOHNSON SPACE CENTER HOUSTON TX

The Houston Space Center NASA's Johnson Space Center Mission Control is the first and the last place that every Space Shuttle Mission visited, NASA and Houston have extensive roots binding them together that goes back to the inception of the United States Manned Spaceflight Program. Mission control was the eyes, ears, shoulders and backs that every shuttle mission was built on. *The Manned Space Flight Center Foundation Inc.* is a 501C non-profit organization. In their eighty two page RFI, the MSFCFI was established in 1986 to further the needs of opening a new visitor center and to show the accomplishments of NASA ad the Houston Space Center. The MSFCFI opened their doors in 1992, and since has been a highly regarded tourist attraction and educational center. The want to acquire a shuttle was backed by the conviction of the MSFCFI and the Johnson Space Center, the Bay area Houston

Partnership with 270 members, engineers, scientists and astronauts, along with elected officials including the Texas Congressional Delegation.

The Center is a $69 million, 183,000 sq. ft. complex that has hosted over 12 million visitors from around the world. They average some 750,000 visitors each year that generated $242 million in total revenue since it opened.

The plans for the orbiter included being connected to the complex adjacent to the Johnson Space Center. Along with the Mercury, Gemini, and Apollo capsules on display and the rockets and SKYLAB trainer, it would add to the classroom facilities for hands on exhibits on space exploration.

The Houston Space Center opened on October 16, 1992 on an 88 acre site just adjacent to Johnson Space Center and is licensed under the MSFCFI until the year 2041.

Because of the sound money management, they rely only on modest fundraising and sponsorship efforts and have been listed on Houston's number one attraction.

Unlike many museums, Houston Space Center does not rely on endowments to fund operations but has been able to generate revenue and update exhibits such as they are needed for repeat business. In short, they have a sound financial background.

With hopes for the Orbiter award, " The space shuttle exhibit is expected to increase Space Center Houston's current $45 million annual regional economic impact. Moreover, the project's construction will generate another $29 million in business value and over 750 jobs. " This quote is from Robert F. Hodgin, Ph.D. University of Houston- Clear Lake Association ,Professor of Economics. Their RFI goes on to expand on their funding sources something the Intrepid RFI did not. In fact, all the other RFI's did go into very detailed funding explanations.

Houston's RFI also went into detailed description of their policies of artifact management. They did not just put a generic collections management policy attached to the end of their fifteen page RFI.

Their educational staff for the center had grown to fourteen full time and twenty part time employees along with seventy volunteers to assist with education efforts. The education department has multiple programs using state of the art teaching methods. They educate nearly 100,000 teachers and students annually from around the world regarding the U.S. Space Program.

Houston Space Center has many corporate led programs that are very successful. The *"BP PHYSICS CHALLENGE"* is in its 4th year and involves over 1200 physics and math high school students. NASA, Boeing, and the Albert and Ethel Herzstein Charitable Foundation and South West airlines are sponsors of the "Annual Space Exploration Educators Conference". In its 15th year, this 3 day event provides some 190 sessions for 580 teachers and informal educators around the country.

Johnson Space Center moved its Educators Resource Center (ERC) to space Center Houston in 1997. This supported free curriculum materials, research tools, videos, books, magazines for teachers designed to enhance math and scientific literacy in students around the U.S. the ERC has served more than 50,000 educators from 1997 to 2008.

The Educational programs listed in the RFI are truly wonderful:
a. Starship Gallery
b. Blast Off
c. Giant screen theater
d. The Feel of Space
e. NASA Tram Tour
f. Astronaut Gallery
g. Mission Status
h. Martian Matrix
i. Kids Space Place
j. International Space Station

They also include:
A. School overnight visits
B. School visits
C. Home School day
D. Orbital Outreach
E. Educator Open House and Johnson Space Center and the Educational Resource Center
F. Virtual Space Community
G. Scout Camp
H. Day Camp
I. Space School

With a massive background of supporter and a superb staff of educators, not to mention the fact that every shuttle mission flown lived and died in that mission control room, there is no logical reason why Houston did not receive an Orbiter.

The US Space and Rocket Center, Huntsville Alabama.

The U.S. Space and Rocket Center Huntsville, Alabama

Located in Huntsville Alabama, the home of the NASA Marshall Space Flight Center, the U.S. Space and Rocket Center in their seventy eight page RFI initially requested the Enterprise Orbiter. The Smithsonian in Washington, D.C. was involved in the discussions regarding the Enterprise and felt that it would have been a good choice. Enterprise would have had to move 696 miles from Washington, D.C. to Huntsville Alabama. It would have allowed them to pay just for the transportation costs. However, they did entertain a chance for a flown orbiter since they hade just completed their Saturn V Housing Project which cost $10.5 million. Funding would have been via the state, donors and corporate community.

In a proposal filed in March of 2009, USSRC asked for the Enterprise because they already had a good display of test artifacts that would have directly related to Enterprise. It would have continued the Marshall Space Flight Center role in the space program history and with out Enterprise, the exhibit would lack in telling an important part of the shuttle history.

Enterprise had a direct connection with Huntsville. In March 1978, OV-101- Enterprise was sent to Huntsville to be mated and to undergo vibration tests. In the months Enterprise spent at Marshall Space Flight Center, the test results gave engineers critical information in designing interface elements. On February 23, 1979, Marshall Space Flight Center finished testing with Enterprise. The Enterprise would have been housed at USSRC in an OPE (Orbiter Protective Enclosure) which in itself is a one of a kind artifact.

USSRC completed a twelve page history of Enterprise' s life with Huntsville. Even Smithsonian was in agreement with the entire concept. In truth, Enterprise seriously belonged to Huntsville and should have been dispatched there and not to a bubble tent on an aircraft carrier, who's space history could not compete with Huntsville, USSRC and the Marshall Space Flight Center.

To keep the record straight, USSRC also had the monetary ways and means to support an Orbiter flown or not. The OPE is climate controlled, 248 feet by 168 feet by 84 feet. It would have a concrete foundation and finished to museum quality standards complete with exhibits, ramps, restrooms and handicap access.

They could meet the delivery date from NASA and funding dates with not problem. They also have the famous *Space Camp* for kids. The NASA Stars program was developed by educational staff at both USSRC and MSFC and was in line with the quality of STEM education in title 1 schools.

There is also a Space Academy for teachers and educators, which gives hands on approach with professional developmental sessions. USSRC has also used Federal grants that involved training students and teachers with NASA Stars Program and NASA Educator Astronaut teachers (NEAT)

The RFI also showed an extensive governing board and provided for their figures for attendance and web page hits. There is no reason why USSRC should not have been considered as a home for the Enterprise. In fact, they should have been the first on the list.

The Tulsa Air and Space Museum in Tulsa Oklahoma

Tulsa Air and Space Museum and Planetarium This museum had the option of using the Tulsa International Airport as a landing site for an Orbiter. In their sixty two page RFI, TASM also maintained a good educational 9STEM) program which allowed kids hands on training
with mission control simulator, space shuttle cockpit simulator and orbiter docking simulator. Boeing Aircraft in Tulsa, built the

Orbiter Cargo bay doors, all eleven of the International Space Station main truss structures and integrated electrical assemblies for the station.

Rockwell International at Tulsa also made the modifications for the Boeing 747 Carrier aircraft to enable it to carry an Orbiter. The shuttle mate, de-mate devices were also made by Rockwell.

TASM had a large contingency of monetary supporters and they were able to meet the delivery date from NASA.

TASM and Oklahoma State University led to :
a. a teachers advisory board composed of educational leader who were tapped to develop and analyze educational programs for TASM
b. Programs measured against PASS (Priority Academic Student Skills), Oklahoma State Board of Education, National Science Objective.
c. Each program evaluated for relevance, student satisfaction and parent satisfaction, attendance records and surveys were kept to assess effectiveness.

TASM too provided a detailed history of its museum, plans for the orbiter, funding, and educational programs along with financial stability. Why was TASM not chosen? The reasoning that California Science Center was twelve miles from where the shuttles were built, actually comes up lame against this museum and its history with the shuttle.

We have now looked at the winners and losers in the shuttle shuffle. It's not hard to discern that there is something screwy with this entire program of NASA selection on who gets an

Orbiter and who doesn't. In viewing each of these RFIs in total, the decisions made regarding their placement makes absolutely no sense at all, at least not to the normal human being . However, the NASA Administrator is not the everyday normal person, he is the head of the U.S. Space Program.

My question is this: who does he answer to, and why was none of the placements questioned? Yes, there were many members of Congress and the American public that did question it, but basically we were all told to get lost. The last time I looked, I thought that Bolden held a government position, which means he works for you and me. He is not a single entity, operating in a vacuum. Bolden should be made to answer for every decision he made and the reasons as to why he made them. If it was up to me, that would be in a Congressional hearing as the why he is still holding his current position.

The Conundrum

In looking at all of the NASA papers regarding the transfer of shuttle artifacts, including the Orbiters, someone needs to explain this fiasco to the American taxpayer. However, before we look at this, we need to see what proceeds it. That would be the *FPAS* Act (Federal Property Assistance Program) This Act developed in 1949 allows for the GSA (General Services Agency) to dispose of excess artifact or properties of government agencies. What this means is that GSA was to help NASA dispose of all of the artifacts created by the Space Shuttle Program. NASA, according to a *Memo of Agreement*, that was signed with the GSA back in 2008 allowed for the GSA to dispose of those properties by offering it to other government agencies, or if there were no takers, then put them up for auction or sale. Anything under the $10,000 limit could be transferred by NASA itself, Anything above $10,000 was to go through GSA. According to the *White Paper on Transfer or Donation of Shuttle Program Hardware to Eligible Recipients (Logistics Mgmt div HQ LMD Dec. 5, 2008)"* Nasa generally does not have the authority to directly transfer its property to other Federal Government Agencies"

Basically what all this comes to is this: NASA was supposed to work not only GSA but other external authorities knowledgeable about the Museum community standards and practices to protect the shuttle's place in history according to *NASA's Personal Property Disposition Plan (Nov. 2008)*. But it ain't how it happened, folks.

Charles Bolden, NASA Administrator, in his own words *"ISOLATED"* his Recommendation Team (the group that made the decisions for the shuttle disposition) from any influence. NASA's Recommendation Team did not include any museum officials! As to the myriad of agreements and papers written to support this fiasco since 2008, no one from any of the museums that deal with either the National Collection, meaning National Air and Space Museum or the American Alliance for Museums (AAM) had anything to do with the decision making regarding the host museums involved in the choice process.

The more you read; the more you can see that the entire process became a convoluted cluster $#@!. I leave the rest to your imagination. Regardless of how many papers NASA published since 2004 when President Bush decided the shuttle program should end after the International Space Station was complete, the main disaster comes down to the RFI (Request for Information). And please note, it took NASA 4 years before they could figure out what to do. In that time period, administrations had changed, along with the administration of NASA.

The **RFI**

The "Request for Information" was the form that a museum filled out and submitted to NASA for review in hopes of receiving an Orbiter. The name itself, is confusing. What is a request for information, What does that say, Request for information regarding what? The title is not clear and neither was the process.

The first RFI that was put out had the following listed on it:
- RFI responses must include:
- Name of primary contact for response
- Academic faculty or business
- Institution or organization affiliation
- Email address
- Phone
- Identification of other key individuals who collaborated on the RFI response
- A brief summary of other key individuals who collaborated on the RFI response
- A brief summary (300 words) on description of previous relevant experience in displaying assets of National significance.

Not really much to go, right? It gets better. NASA went on to request responses to the following questions:
- Would your organization be interested in acquiring an Orbiter/ SSME (space shuttle main engine). For what purpose and at what location?

- Please explain your organizations approach to raising funding necessary for Orbiter safing and final display preparations, SSME assembly and final display preparation and transportation service.:
- A. What would be the proposed source of funding?
- B. what is the estimated amount of time needed to raise sufficient funds to display an Orbiter/SSME?

- What space shuttle orbiter and SSMEs may not be displayed outdoors and will require suitable climate controlled indoor display space. Please provide your organization's capability to approximately house, protect, display and curate an Orbiter or SSME?
- Given the financial & curatorial requirements stated in their RFI, what is the earliest date your organization could accept the transfer of an orbiter of SSME?
- What is the benefit to the nation of displaying a space shuttle orbiter or SSME at your facility? In your response, please identify:
- A. How you would use these3 assets to inspire the American public/ students in particular?
- B. Other specific educational or education outreach opportunities and
- C. How would you assess, evaluate, and measure these objectives.?
- Provide the techniques and interpretive strategies that you would use to enhance the display of the artifacts and increase the public's ability to understand the Nation's space exploration agenda.
- What added assets, tools or expertise would your organization need from NASA in order to display these assets to the public?

- Topics which organizations should also include the following in an Appendix in as much detail as reasonably possible:
a. Mission statement
b. Organizational Chart
c. Nature of governing authority
d. Accreditation or other relevant credentials
e. Collection ownership and management policy
f. Attendance figures for the past 5 years.
g. Population of geographic area in which your organization is located.
h. Local infrastructure for transporting an orbiter once offloaded from shuttle carrier aircraft to display location
i. Budget resources profile including endowments over past 5 years.
j. Number of webpage hits for each of the past 5 years.

Please note the RFI responses including the appendix should not exceed twenty five pages in length. Use since space 12 point Times New Roman font. Use a PDF or doc file.

Looking at these questions, one would think that they are not extremely difficult[13]. They are for all purposes yes and no or an explanation. So, one would think answering them wouldn't be too hard. Question then why were there so many mistakes on the part of NASA in reviewing and posting the information to a matrix? Yes, there were mistakes like the Intrepid claim that they were accredited by the American Alliance for Museums, which is a total lie. Another question would be why the National Museum of the U.S. Air Force was shown as not accredited and shown to be in the West instead of the Mid-West. The list of errors goes on and they are cited in the OIG

---

[13] *Initially looking at the RFI, I have answered deeper questions regarding my household in adopting a pet. While a life is much more important, still when it comes to tax payer dollars and our history, I would think a more precise method could have been used instead of a 3 page 300 word composition about what you would do if you got a shuttle.*

report on the Shuttle Orbiter Disposition report.

Yet there is not one word on how these museums would upkeep and maintain the orbiters. Wear and tear on objects on display happens due to environment, and just being. They will not stay pristine forever. Who, what and where they get the knowledge and equipment is a question. There is also the question of title. How does NASA give a legal title to a museum when this is tax-payer property? It's a question not asked, not answered.

There is also the question of why there are three shuttles on the eastern seaboard and one on the west coast, California. Letters of complaint from Congressional parties flew into NASA's Administrator's office when the decision finally came down. The largest hit was Houston, Texas, home of the Johnson Manned Spacecraft Center, the controller of all shuttle missions since 1982. How could Administrator Bolden and his "Recommendation Team" bypass the Johnson Space Center and not give them a shuttle? The concept is truly ludicrous. In a letter[14] written by Congressmen Al Green and Congresswoman Sheila Jackson-Lee, it was stated how "disappointed" they were in not receiving an Orbiter. They went so far as to suggest that NASM (National Air and Space Museum) lend them a shuttle for a term of 15 years. Of course, it was an act of desperation on the Congressional Representatives part. Bolden explained how it would not be possible since the equipment and SCA (Shuttle Carrier Aircraft) would not be available and not held in abeyance to service the shuttle after the

---

[14] *This letter will appear in the appendix*

program shut down and the Orbiters reached their destination. The only thing that Houston finally did get was a "REPLICA" of a shuttle. Can you believe it? I guess that NASA finally had to acquiesce to the bad publicity and had over something to Houston. The best NASA could do was a replica.

This shuttle replica was delivered via barge to Galveston Bay under a fireworks display. Really sort of an understatement when you consider that Enterprise which should have been in Houston is sitting in a bubble tent on Intrepid. (Photo NASA website)

In another letter[15] released under the FOIA act, the entire Texas Congressional team wrote a letter to Administrator Bolden demanding to know why Intrepid Sea Air space Museum was considered a better bet than they were . The letter states:
a. *What factors did you use in making your decision?*
b. *For what specific reason was Intrepid in NYC chosen?*
c. *Are there any historical connections between NASA and Intrepid?*
d. *Are there any historical connections between NASA and NYC in general.*

---

[15] This letter will appear in the appendix

> e. Exactly how much does NASA anticipate spending to move the Enterprise from its current location in Virginia to Intrepid?
> f. Specifically what funds will NASA use to pay for this move?
> g. How does NASA physically plan to move the Enterprise to Pier 86 in NYC?
> h. Considering the Intrepid is located only 224 miles from the NASM Udvar-Hazy Center in Virginia, where the shuttle Discovery will be displayed, doesn't it make geographic sense to have Enterprise more centrally located in the country.

"As it stands now, there are three location on the east coast. Wouldn't a more central location ensure that the highest number of Americans would be able to visit the orbiter?"

The letter went on to vehemently state its hope that politics had not played a part in this historic decision and there is no rational explanation based on definable factors for the choice of Intrepid in NYC and the transfer of enterprise to that location will cost significantly more than a transfer to Houston. The Congressional team went on to say " We will do everything in our power in congress, including legislation to prevent funding of the transfer to stop the wasteful decision." The say the least, the vitriol was roiling in Texas.

Administrator Bolden's response[16], which was also part of the FOIA packet, to this very angry letter was the usual "Bed Bug" type of response, which cited "good things about his decision". For those of you not familiar with" Bed Bug letters", they are a form of response usually written by government or military offices which answer a tax payer's letter on an issue that to the particular government office is

---

[16] Bolden's response in full will appear in appendix

not of consequence, but must be officially answered in a polite and unobtrusive, manner. The letter just goes to show Bolden's arrogance regarding "his" decisions as final on the Orbiters disposition.

To further show how political lobbying did play a part in orbiter placement, another letter from the FOIA request involved Senator Charles Schumer of New York City. The letter was addressed to Administrator Bolden. It requested that the administrator please consider using Stewart Airport in New Windsor, Orange County, New York, to land the shuttle. In this letter Senator Schumer listed all the benefits and financial gain garnered if the shuttle were to land here instead of JFK International Airport. In the long run, Schumer's request cited how Henry Hudson had traveled up the Hudson river 400 years ago and how wonderful it would be to show off the shuttle traveling down river to its new home. That did happen anyway, but not the way Schumer wanted.

The next letter released in the FOIA packet comes in the request from an outraged Ohio House of Representatives requesting the cost and justification of the decision to hand shuttles everywhere but the National Museum of the US Air Force. They also went so far as to request a shuttle on loan. The FOIA also brought out the snowballing of NASA by sending in that packet of "releasable material" at least twenty form letters, sent to various Congressional individuals, letting them know of the final shuttle decision.

In the letter received with the releasable material, it was stated that another re-review of the 2161 documents (most of them duplicated) will be done and we would be notified by August 17, 2012 as to the determination. The letter received with the material also stated that NASA would now ask the museums' involved if they would release the financial reports that were submitted in the RFI. They will have a determined time with which to respond. We will wait for that, too.

The amount of material regarding this entire scenario is amazingly lacking in clarity, as least in the FOIA request material. Should you look online, however, you will find scads of reports and documents which NASA has put out regarding the entire situation. The problem here is, there is so much information that is conflicting in each report put out that it really takes a lot of time and effort to figure out what is right and wrong. The other part of the problem with making sense out of the shuttle shuffle has to do with NASA itself. In order to even get close to finding out how NASA interpreted the law of the land, this author went through many phone calls, being shuffled to departments that had nothing to do with the questions being asked, and in essence stonewalled. As of right now, there is no answer to the many questions that I have asked regarding the decision making process and the information that is posted on the NASA website.

The White Paper

One of the documents that you can find online is the *"White Paper Transfer or Donation of Shuttle Program Hardware to Eligible Recipients"*[17]
The paragraph below is from this paper:

---

[17] http://www.hq.nasa.gov/oia/nasaonly/itransition/Transfer-Donation.pdf

"The GSA was established by the Federal Property and Administrative Services Act of 1949 (FPAS). The purpose of the Property Act was to simplify the procurement, use, and the disposal of Government property and to assure that public property is fully used for public purposes, prior to its sale or disposal. The Property Act assigns the GSA Administrator the responsibility for the supervision and direction over the disposition of excess and surplus Federal property. The Space Act also provides disposition instructions, requiring NASA to disposition property according to FPAS. Specific details of GSA's support to NASA, including reimbursable rates for Shuttle excess hardware, and NASA's responsibilities to GSA are outlined in an MOA, signed between GSA Assistant Commissioner Joseph Jeu and NASA Assistant Administrator Olga Dominguez, dated February 20, 2008."

However, that isn't what actually happened. In all of the papers that are accessible on line via the NASA website regarding the shuttle disposition, they all say the same thing. GSA should have been responsible for the vetting of museums in the Orbiter disposition process. The question of how and where Administrator Bolden took it upon himself to decide how the Orbiters would be distributed. Bolden decided that it was more urgent that the international and local populace see the Orbiters, than to place them where they actually might have done the most good, and be protected as national treasures.

The White Paper makes it abundantly clear that NASA should be working along the lines of FPAS (Federal Property Assistance Act) which was created in 1949 by Congress with enactment of Public Law 94-519. this law enables federal excess property to be donated to agencies that meet the requirements. The next part of the statement leads into the GSA (*General Services Administration*) which helps federal agencies to dispose of excess material or artifacts. Basically, NASA and the Bolden Administration swept that whole section under the carpet and decided to override what the law of the land says.

American Alliance for Museums Standards

The American Alliance for Museums[18] (formerly the American Association of Museums) is the cornerstone of the museum industry. In its many years of service, it has conceived a code of ethics that responsible and reliable museums follow. Its time to show just what NASA missed, by not employing this agency in respect to the disposition of the orbiters. The AAM makes the statement the " ethical codes evolve in response to changing conditions, values and ideas." Be assured that the AAM is up to date in regards to the needs of today's museum system.

AAM states that museums make their contribution to the public by collecting and preserving and interpreting things of this world. This is done to ""advance knowledge and nourish the human spirit." A museum is not in the business of donating or accepting artifacts based on general populace or location. The AAM goes on to state that museums in the United States are grounded in a tradition of public service. They are "organized as public trusts, holding their collections as a benefit for those they were established to serve." The law

---

[18] *Recently changed their title after 100 years of service to American Alliance for Museums from American Association of Museums (AAM)*

provides a museum's basic framework for museum operations. Because they are non-profit institutions, all museums must comply with the 501-C law which exempts them from paying taxes. There are local, state and federal laws that need to be complied with, as well as legal responsibilities.

As far as collections go, each museum is driven by its collections. The stewardship of these collections necessitate the " highest public trust and the presumption of ownership, permanent care, documentation, accessibility and responsible disposal."

A museum ensures that:[19]
a. collections in its custody support its mission and public trust responsibilities
b. collections in its custody are lawfully held, protected, secure and unencumbered, cared for and preserved.
c. Collections is custody are accounted for and documented.
d. Access to the collection and related information is permitted and regulated.
e. Acquisition, disposal and loan activities are conducted in a manner that respects the protection and preservation of natural and cultural resources and discourages illicit trade in such materials.
f. Acquisition, disposal and loan activities conform to its mission and public trust responsibilities.
g. Disposal of collection through sales, trade, or research activities are solely for advancement of museum's mission.
h. Collections related activities are to promote the public good rather than the individual financial gain.

---

[19] http://www.aam-us.org/museumresources/ethics/coe.cfm

i. Competing claims or ownership that may be asserted in connection with objects in its custody should be handled openly, seriously, responsibly and with a respect for the dignity of all parties involved.

Here we see a partial list of the "Ethics" that a good museum should uphold. It also shows how a museum should operate and be treated by society, that meaning the business world and other museums. When you look at how the orbiters were acquired, it's fairly obvious that none of these rules were applied to that process. In fact, when you look at what Administrator Bolden proposed with his "Recommendation Team", the isolation he imposed on them, along with the lack of museum expertise on the team and his personal
ideas injected into that has led to a disastrous situation in regards to the placement of national treasures. As we have shown in our breakdown of the museums that did receive an orbiter, the best choice was not made, The worst choice possible was made because Bolden and his team did not adhere to basic and fundamental museum practice.

If you ask the question why didn't the museums involved protest, that is an easy question. THEY DID! First, The Intrepid will take anything that will bring them more business. The weak connection between NASA and Intrepid was not enough to support them receiving an orbiter. The museum has had numerous problems in managing their collection. While in 2010 they did adopt a Collections Program, it is a simple document and doesn't stand to what and how Intrepid portrays itself to the public. It is always on the lookout for the next " Activity" or "Day" as explained before. There are many museums that have "Activity" days or special program days for kids and such. There is nothing wrong with that. But to be sure, these museums need to draw the public because they are not getting $30 per adult and $23 for kids to see a shuttle in a balloon tent. If Intrepid was true to its cause, it would have realized that a shuttle in a tent in not the optimum place to show off an orbiter, be it atmospheric or flown in space. Intrepid cared nothing for the Orbiter only what the Orbiter would afford them. If you think I am being harsh in this judgment. I have seen much regarding this museum and for a NASA Administrator to crap in the face of the birthplace of aviation, namely Ohio, to give an Orbiter to a ship that has NO substantial history with the space program, forgive me. I am a historian and I protect history, I don't use it to pay bills. How a NASA Administrator can turn on his own, meaning Houston, the mother of all shuttle missions, is beyond the scope of reason. This all goes back to the statement that Bolden did not adhere to museum practice and ethics in placing the Orbiters.

Solution and Conclusion

Ladies and Gentlemen, we come to the end of this criticism. What purpose did it serve? I will tell you. It made you aware as a taxpayer, that your money and your history are being squandered. Your history is placed without thought or care into places that it shouldn't be, by a bureaucrat who had only one thing on his mind, get rid of the shuttles and place them where he and he alone felt it would do the most good. This is a democratic society, at least it was the last time I looked and when it comes to our history, I believe that our lawmakers, who went to bat for their states in getting an orbiter were literally kick in the nether regions by one man's decision making process. To add to this picture of contempt for the space program, just about every day I receive notices from NASAs's RSS feed about the next item that they are quickly disposing of, like the Liberty Star and Freedom Star. These are the vessels that picked up and returned the SRBs to Cape Canaveral to be reprocessed for the next shuttle flight. Liberty[20] is now being shuffled off to a Maritime Academy as a training vessel in Kings Point ,New York.

Yet, the best that I have heard is that NASA has now selected *Barrios Technology Ltd. of Houston* to now provide the mission and program integration services to handle the strategic and tactical planning, engineering analysis, manifest development and hardware certification, safety and mission assurance, risk management, information technology, payload integration and program science and research for the ISS. What happened to all the NASA people in Houston, highly trained NASA people of Houston, that now have been laid off. This number adds into the 13,000 that were laid off in Houston, soon as the program closed down. It adds up to this, not only are we not controlling our part of the ISS via NASA engineers, we now have the lowest bidder with a two year contract doing the job. You ask why this paper was written?! It's just the tip of the iceberg that has begun to sink the Manned Spaceflight Program for the U.S.

This debacle is courtesy of the Obama Administration, folks. We are losing our best and brightest. I am not against business and the commercial sector getting some work. I am

---

[20] these documents regarding disposal of Liberty and the contract for the ISS appear in the appendix.

against the loss of technical aptitude and experience we won't ever get back. Everyday, little "cherries" like this are released from NASA regarding the remains of the Shuttle program. I question *everything* that the Obama administration has enforced on NASA and I question Charles Bolden and his administration intensely. So should you, since your taxpayer dollars paid for all of the equipment they are divesting themselves of.

The process managed by Bolden can't be allowed to stand. The lawmakers, who in fact, did rally for their states, and the taxpayers who foot the bill should demand the removal of Administrator Bolden from office. We can't do anything about it, you say? Yes, you can!! Email your senators and your Congress men and women. Tell them that you will not tolerate the use of your history and money in this manner. Explain that this is **YOUR** dime and you are entitled to have someone at NASA who knows and cares about the United States and its place in the manned space program. Don't pay $30 to see an orbiter in a balloon tent and tell the museum why. Can we move the Orbiters? It can be done, carefully and with forethought to the significance of the Orbiter. Don't allow a bureaucrat to get away with stealing your history and money. Don't allow museums like the National Museum of the US Air Force and the Seattle Museum of Flight to be ignored, tell your Senator or Congress men and women how you feel. If you allow this to continue, it won't be long before many of our national treasures are just given to the highest bidder or the more politically correct venue.

My solution to this distressing situation is a short and easy one. Put the SCA back into service, revisit each museum and their RFI and have someone with some sense of history and museum practice like NASM and the AAM sort this disaster out. KEEP NASA'S FRONT OFFICE OUT OF IT !! There might be a chance to save the Enterprise and Endeavor from fates of fatigue and lack of maintenance and even the score out as to three shuttles on the east coast and one on the west coast. I believe this can be done without much more cost being added to the process, and with this being weighed against the loss of an orbiter due to poor care and unsafe conditions, the cost would be negligible.

Remember, those who don't read history are condemned to repeat it. This scenario was nothing more than the political strategy of a bureaucrat in a position of power. Let's not let it happen again. Let's hope we can turn this use of our history to something more prestigious than a $30 ticket and an object that sits so it can be used as a means of status instead of an educational exhibit that will last generations for our children and their children.

We paid for this program. We should have our money's worth. We should also demand that the parties involved with this despicable situation be removed from their office. They have no place in and do not deserve the power and prestige that goes with their office. My only hope is that it may not be too late to really do something about this mess, but *politics* being what they are, I would not expect much to change, not unless there is a public outcry. It's up to you American Taxpayer, save your history now or be prepared to lose it forever.

Appendix

The *Request for Information* for each of the responding museums can be found at our author's webpage on Amazon:
http://www.amazon.com/s/ref=nb_sb_noss?url=search-alias%3Dstripbooks&field-keywords=jeannette%20remak&sprefix=jeannette%20re&rh=n%3A283155%2Ck%3Ajeannette%20remak

Unfortunately, due to the large number of pages involved with these RFIs (over 500), it was not possible to include them in this publication. However, we did manage to post them on KINDLE without the book attached so that you can read just the RFIs. The charge is $1.99. We didn't want to charge anything but KINDLE is not set up to do it for free. The proceeds for the RFIs will be donated to *Wounded Warriors Foundation* because we will not take money for something that belongs to the American people via the Freedom of Information Act.

The document below is notice of how NASA transferred or gave ownership to the various museums. The only issue with this is since NASA did not own the shuttles outright as they belong to the American people, there should have been a permanent loan agreement, not a transfer of TITLE.

> 9/29/12
> Welcome to English.news.cn
>
> NASA transfers Endeavour's ownership to Californian museum
>
> **NASA transfers Endeavour's ownership to Californian museum**
>
> LOS ANGELES, Oct. 11 (Xinhua) -- U.S. space agency NASA on Tuesday handed over the ownership of retired space shuttle Endeavour to the California Science Center in Los Angeles, where it will go on permanent display.
>
> "NASA is pleased to share this wonderful orbiter with the California Science Center to help inspire a new generation of explorers," NASA Administrator Charles Bolden said in a statement.
>
> "The next chapter in space exploration begins now, and we're standing on the shoulders of the men and women of the shuttle program to reach farther into the solar system," he said.
>
> According to NASA, Endeavour is expected to arrive at Los Angeles International Airport next year and then be towed through the streets to its new home in California, the state where it was built more than 20 years ago.
>
> "Endeavour now will begin its new mission to stimulate an interest in science and engineering in future generations at the science center," said Jeffrey Rudolph, president of the California Science Center, during a ceremony at the museum.
>
> Endeavour, which flew its final voyage in May 2011, had traveled 115 million miles (184 million km) during 25 flights and carried 139 people into orbit before NASA retired the space shuttle fleet in July.
>
> At the center, the retired orbiter will be housed horizontally at first until a permanent exhibit is built. Its final position will be vertical, just as if it were ready to launch.
>
> The value of the space shuttle is estimated at 2 billion U.S. dollars. And the total cost of transportation, as well as building a new wing at the museum, is expected to reach 200 million dollars.
>
> The logistics and funds are still being hammered out. The museum has already received 20 million dollars in promised donations, according to ABC.
>
> Endeavour was built from 1987 to 1991 to replace the destroyed Challenger shuttle, which exploded 73 seconds after its takeoff on Jan. 28, 1986.
>
> The California Science Center is one of the four locations across the country selected to display NASA's retired space shuttle vehicles. Museums in suburban Washington, D.C., Florida and New York will receive the remaining shuttles and a prototype vehicle.
>
> Copyright 2010 Xinhua News Agency

This is a web article which describes the Houston, Texas desperate want for an
Orbiter.

An article about the Hopes for the Houston delegation to get an Orbiter.

Houston delegation wants a new shot at shuttle

By STEWART M. POWELL, HOUSTON CHRONICLE | Thursday, September 29, 2011 | Updated: Friday, September 30, 2011 12:37am

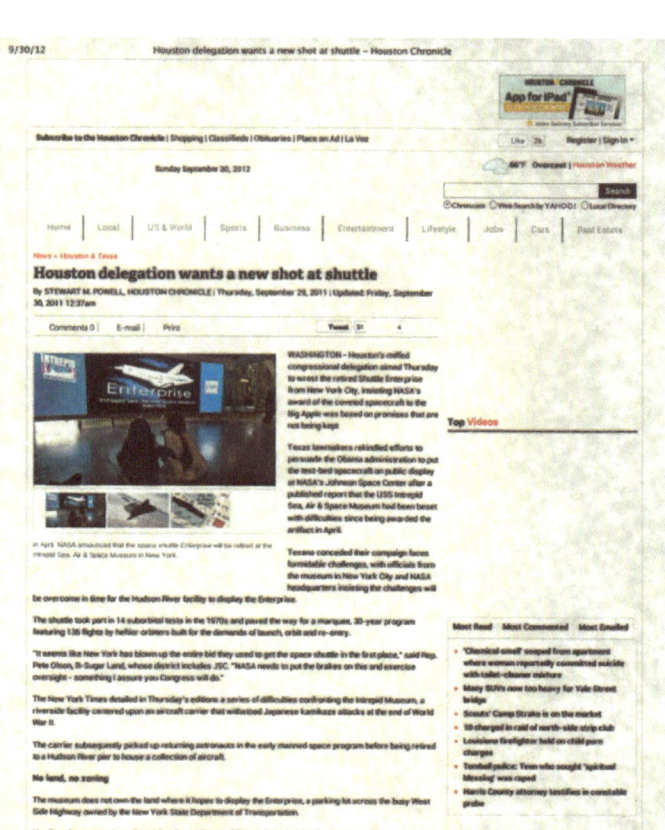

In April, NASA announced that the space shuttle Enterprise will be retired at the Intrepid Sea, Air & Space Museum in New York.

WASHINGTON – Houston's miffed congressional delegation aimed Thursday to wrest the retired Shuttle Enterprise from New York City, insisting NASA's award of the coveted spacecraft to the Big Apple was based on promises that are not being kept.

Texas lawmakers rekindled efforts to persuade the Obama administration to put the test-bed spacecraft on public display at NASA's Johnson Space Center after a published report that the USS Intrepid Sea, Air & Space Museum had been beset with difficulties since being awarded the artifact in April.

Texans conceded their campaign faces formidable challenges, with officials from the museum in New York City and NASA headquarters insisting the challenges will be overcome in time for the Hudson River facility to display the Enterprise.

The shuttle took part in 14 suborbital tests in the 1970s and paved the way for a marquee, 30-year program featuring 135 flights by heftier orbiters built for the demands of launch, orbit and re-entry.

"It seems like New York has blown up the entire bid they used to get the space shuttle in the first place," said Rep. Pete Olson, R-Sugar Land, whose district includes JSC. "NASA needs to put the brakes on this and exercise oversight – something I assure you Congress will do."

The New York Times detailed in Thursday's editions a series of difficulties confronting the Intrepid Museum, a riverside facility centered upon an aircraft carrier that withstood Japanese kamikaze attacks at the end of World War II.

The carrier subsequently picked up returning astronauts in the early manned space program before being retired to a Hudson River pier to house a collection of aircraft.

No land, no zoning

The museum does not own the land where it hopes to display the Enterprise, a parking lot across the busy West Side Highway owned by the New York State Department of Transportation.

Nor does the museum have the zoning change that would be needed to build and operate a museum on land reserved for industrial manufacturing.

This is a letter from the Ohio House of Representatives detailing its case in not receiving an Orbiter for the National Museum of the United States Air Force.

Ohio House of Representatives

RECEIVED
EXECUTIVE SECRETARIAT
2011 APR 28 A 7 31

April 18, 2011

NASA Administrator Charles Bolden
NASA Headquarters
Suite 5K39
Washington, DC 20546-0001

Dear NASA Administrator Charles Bolden,

On April 12, 2011 it was announced that NASA would not be donating a retired space shuttle to the National Museum of the U.S. Air Force at Wright-Patterson Air Force Base. We would like to register how deeply disappointed we are that the executive committee failed to recognize the tremendous advantages of housing an orbiter at Wright-Patterson. We would also like to voice our concern with the selection process, which offered little transparency or explanation of the final decision.

Housing an orbiter in Southwest Ohio would have been an enormous boon for the region, bringing with it potentially millions in revenue and tourism. Wright-Patterson Air Force base is within one day's drive of 60 percent of the U.S. population and already draws over a million visitors within a year's time. It is home to exceptional science and educational programs as well an already impressive display dedicated to NASA. Also, Wright-Patterson Air Force Base has the capabilities in place to house a shuttle, which would include a climate controlled setting. Having a retired space shuttle here would have been a great opportunity to an area that already has much aviation and aeronautical history.

We respectfully request a review and cost justification regarding this decision. We are also asking for the possibility to house a shuttle on a temporary basis.

Sincerely,

Peter Beck
67th House District

Cliff Rosenberger
86th District

Terrance Blair
38th District

Jim Butler
37th District

Jarrod Martin
70th District

Michael Henne
36th District

cc: Governor John Kasich
Speaker William G. Batchelder

www.house.state.oh.us
77 S. High Street, Columbus, Ohio 43215-6111

This letter to Charles Bolden, NASA Administrator from Senator Chuck Schumer of New York asking for the Orbiter to be brought to another airport instead of JFK to "further" his constituents in that area need for publicity.

CHARLES E. SCHUMER
NEW YORK

United States Senate
WASHINGTON, DC 20510

April 14, 2011

The Honorable Charles F. Bolden, Jr.
Administrator
National Aeronautics and Space Administration
300 E Street, South West
Washington, DC 20546

Dear Administrator Bolden:

  I write regarding the recent announcement that the Intrepid Sea, Air & Space Museum will receive the Space Shuttle Enterprise after the NASA shuttle program comes to its end. Thank you for choosing New York City as the future home of one of America's greatest examples of technological innovation and pride. I have no doubt that the Enterprise will be put to great use in New York, inspiring the next generation of scientists, pilots, explorers and engineers to dream big and to believe that not even the sky is the limit.

  As you are aware, the transport of each shuttle can be undertaken by the Shuttle Carrier Aircraft, which enables an easy voyage for the shuttle vehicle over long distances. It is my understanding that each shuttle to be transported would be mounted on top of this modified Boeing 747. As the planning commences on Enterprise's final journey, I urge NASA to land the shuttle at nearby Stewart Airport in New Windsor, Orange County. Stewart Airport, which continues to serve as a military airfield as well, is home to the 105th Airlift Wing of the New York Air National Guard and Marine Aerial Refueler Transport Squadron 452 United States Marine Corps Reserve, and has one of the longest runways available anywhere in the nation. A fact that qualifies it as one of the alternative landing locations for actual space shuttle missions. Moreover, its proximity to New York City and the Hudson River make it the perfect place to land the Enterprise.

  Floridians, Californians, and New Mexicans have all been able to witness a shuttle landing, but that opportunity has never been available in the New York vicinity. Bringing the shuttle to Stewart Airport would attract visitors from around New York and the whole northeast. With such a large potential for tourism, landing the Enterprise at Stewart Airport would provide a major economic boost to the New Windsor area. Additionally, Stewart Airport is located close to the Hudson River, which would enable the shuttle to be easily transported to a barge. From there the barge can travel down the Hudson River to its final location on Manhattan's west side, making the Enterprise available for viewing during its voyage down the Hudson River. In the days of yore, Henry Hudson first sailed up the Hudson River in a wooden boat. What a fitting testament to America's progress it would be to see the fruit of America's accomplishments being transported down that same river over 400 years later. The landing and transport would be a spectacle and would draw maximum attention to NASA and America's space program.

Thank you for your attention to this request. Please don't hesitate to contact my Washington, D.C. office with any questions.

Sincerely,

Charles E. Schumer
United States Senator

This is an article from the Springfield Ohio Daily Sun discussing how the whole shuttle disposition process was kept very quiet and little was revealed to not only Congress but the
public.

This is a response from NASA administrator Bolden to Texas Representative Sheila Jackson Lee regarding the Texas RFI and why they did not receive a shuttle

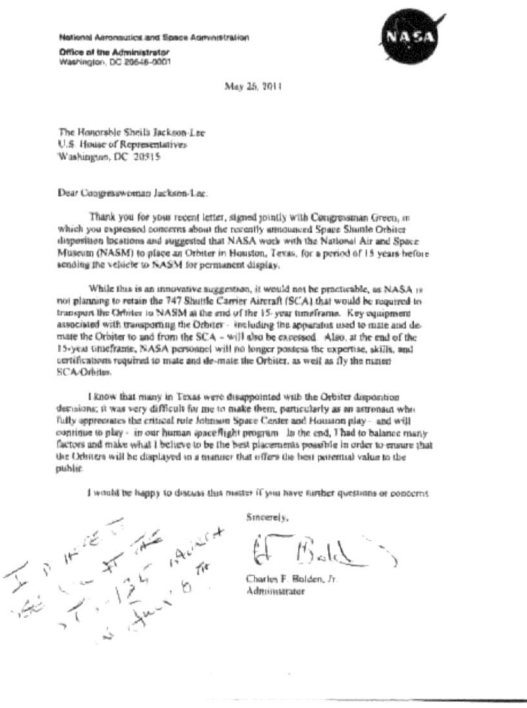

This letter is also to Texas representative Al Green regarding why Houston didn't receive an Orbiter..

National Aeronautics and Space Administration
Office of the Administrator
Washington, DC 20546-0001

May 26, 2011

The Honorable Al Green
U.S. House of Representatives
Washington, DC 20515

Dear Congressman Green:

Thank you for your recent letter, signed jointly with Congresswoman Jackson-Lee, in which you expressed concerns about the recently announced Space Shuttle Orbiter disposition locations and suggested that NASA work with the National Air and Space Museum (NASM) to place an Orbiter in Houston, Texas, for a period of 15 years before sending the vehicle to NASM for permanent display.

While this is an innovative suggestion, it would not be practicable, as NASA is not planning to retain the 747 Shuttle Carrier Aircraft (SCA) that would be required to transport the Orbiter to NASM at the end of the 15-year timeframe. Key equipment associated with transporting the Orbiter – including the apparatus used to mate and de-mate the Orbiter to and from the SCA – will also be excessed. Also, at the end of the 15-year timeframe, NASA personnel will no longer possess the expertise, skills, and certifications required to mate and de-mate the Orbiter, as well as fly the mated SCA/Orbiter.

I know that many in Texas were disappointed with the Orbiter disposition decisions; it was very difficult for me to make them, particularly as an astronaut who fully appreciates the critical role Johnson Space Center and Houston play – and will continue to play – in our human spaceflight program. In the end, I had to balance many factors and make what I believe to be the best placements possible in order to ensure that the Orbiters will be displayed in a manner that offers the best potential value to the public.

I would be happy to discuss this matter if you have further questions or concerns.

Sincerely,

Charles F. Bolden, Jr.
Administrator

The letter that was sent by Representatives Lee and Green to Administrator Bolden regarding why they did not receive an Orbiter.

Congress of the United States
Washington, DC 20515

April 14, 2011

Major General Charles F. Bolden, Jr
Administrator of the National Aeronautics and Space Administration
National Aeronautics and Space Administration Headquarters
Washington, DC 20546-0001

Re: Proposal for Retirement Locations of NASA's Space Shuttle Orbiters

Dear Administrator Bolden:

We are profoundly disappointed with the decision to overlook the heart of America's spaceflight program, the Johnson Space Center, which also serves as the home for the astronaut corps. Our city, our region and our state are deeply invested in the past and the future of the human spaceflight program, and would offer a Space Shuttle Orbiter a place where it would truly be appreciated by the public as well as the workforce which has had a vital role in making the human spaceflight program a success.

As a matter of fair reconsideration and an acknowledgement of Houston's vital role in the history of space flight, we respectfully request that the Space Shuttle Orbiter be made available to, and situated in Houston on a loan basis from NASA for a period of 15 years, after which it would be returned to its location in the Smithsonian's National Air and Space Museum.

We believe the decision to deny Johnson Space Center, the home of Mission Control, a space shuttle is unjust and should be reconsidered. This decision when coupled with the earlier decision that cuts approximately 15,000 Johnson Space Center jobs adds the insult of losing a shuttle to the injury of losing jobs.

We strongly urge your reconsideration of both decisions as they collectively deny our region's importance as an integral part of the NASA team and will adversely impact our economy. We believe that the above proposal offers a fair and equitable opportunity for NASA, the Smithsonian and Houston to all share in this tribute in America's legacy of space flight. We look forward to your response, as the consequences of these decisions for our community will be significant.

Sincerely,

AL GREEN
Member of Congress

SHEILA JACKSON-LEE
Member of Congress

Every Representative of the Texas delegation received this exact same letter regarding the demise of their hopes for an Orbiter. The author still can't see how you could equate the history that the Johnson Space Center has with the Intrepid.

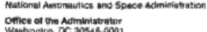

National Aeronautics and Space Administration
Office of the Administrator
Washington, DC 20546-0001

May 26, 2011

The Honorable Pete G. Olson
U.S. House of Representatives
Washington, DC 20515

Dear Congressman Olson:

Thank you for your recent letter, signed jointly with other Members of the Texas delegation, in which you expressed concern about the recently announced Space Shuttle Orbiter disposition locations. It was difficult for me to make these decisions. There were many worthy candidates, including Space Center Houston, each of which had much to offer in terms of providing a good home for our retiring Shuttles. In the end, I had to balance many factors and made what I believed to be the best placements possible in order to ensure that the Orbiters will be displayed in a manner that offers the best potential value to the public.

Addressing one of the concerns in your letter, I want to assure you that political considerations played absolutely no role in my decisions. The process I used was consistent with the NASA Space Shuttle Property Disposition Plan of 2008 and NASA Authorization Acts of 2008 and 2010, respectively, and informed by responses to the two NASA Requests for Information (RFIs), as well as by additional research. Data gleaned from these sources included, but was not limited to: financial aspects of transfer process; quality and availability of facilities; options for, and feasibility of, transporting Orbiters; attendance levels at prospective recipient organizations; size of regional population; and access to international transportation. There was no single criterion used to determine the ultimate selection of recipients.

The Intrepid Sea, Air, and Space Museum, along with the other recipient locations, had an appropriate mix of characteristics, including a connection to human spaceflight. USS Intrepid was the primary recovery ship for the first crewed Gemini mission (Gemini 3, flown by astronauts John Young, who went on to become the first Space Shuttle Mission Commander of STS-1, and Gus Grissom) and Scott Carpenter's Mercury mission. The Grumman Corporation built the Apollo Lunar Module at its facility in nearby Bethpage, NY. As with the other recipient organizations, Intrepid offers a venue that ensures the Orbiter will be visited by large numbers of American citizens

75

While the details of transporting Enterprise to the Intrepid Museum must be worked out with the recipient, it should be noted that NASA is not paying for the display preparation or ferry costs of three of the Space Shuttle Orbiters. The only exception is Discovery, which is going to the Smithsonian's National Air and Space Museum, the curator of the national collection. In the case of Enterprise, there are no display preparation costs as it is currently on public display. As stipulated in the RFIs, recipient organizations must provide funding for the reimbursable work NASA will perform to ferry the Orbiters to suitable airports near the receiving organizations at an estimated cost of $8 million.

Ground transportation options from the airport to the ultimate display location, and therefore costs, vary from location to location. NASA will work with the recipients to coordinate the safe ground transportation of the Orbiters to their respective destinations, which will include on-site visits to each location. In the specific case of Intrepid, ground transportation options could include towing the Orbiter via existing roads and/or barging it via local waterways to the new structure being built to house the Orbiter. As with the air transportation "ferry" costs, recipients are responsible for paying all ground transportation costs.

I want to emphasize again what a challenge these decisions were to make. As a former astronaut, I appreciate more than most the critical role that the NASA Johnson Space Center and Houston continue to play in our Nation's human spaceflight program, particularly as we move into the next decade of International Space Station operations and look toward venturing beyond low-Earth orbit.

It is important that all Americans have the chance to see first-hand one of the great achievements in our space history, and I believe the placement of the Orbiters offers the best value to the American public as a whole.

I would be pleased to discuss this matter with you in greater detail if you wish.

Sincerely,

Charles F. Bolden, Jr.
Administrator

Congress of the United States
Washington, DC 20515

February 14, 2012

Dear Administrator Bolden:

Your response to the October 13th letter sent by me and 40 of my House colleagues raises far more questions than it answers. In your response, you assert that the Intrepid Sea, Air and Space Museum originally proposed "a new museum either on the pier, on a barge, or on land adjacent to the museum." This contradicts all public statements made by Intrepid Museum officials in the months leading up to the shuttle location announcement on April 12, 2011. On multiple occasions, Intrepid Museum officials described plans to house the shuttle in a glass enclosure either on the Intrepid itself or on the adjacent pier. There was never any mention of a plan to display the shuttle in a facility that must be constructed on an empty parking lot across the street on property the Intrepid Museum does not currently own.

American taxpayers funded every aspect of the Shuttle Program and they deserve full transparency and accountability with respect to how the Orbiters will be cared for. We hereby request that you provide us with a copy of the Intrepid Museum's original design proposal, un-redacted, as submitted to NASA.

Furthermore, in your response to our letter, you refer to an April 2011 meeting NASA held at Kennedy Space Center with representatives of each Orbiter recipient museums regarding drafts of their display concepts and exhibit engineering concepts. We request all of the minutes of that three-day meeting. This includes the referenced "logistics plans, exhibit plans and finance plans" for the Intrepid Museum's acquisition of the *Enterprise*. NASA's decision on which entities would host the Orbiters was made prior to April 2011 (as confirmed in the NASA IG report *Review of NASA's Selection of Display Locations for the Space Shuttle Orbiters*). There is no reason why these minutes should not be shared with Congress and the public.

Your response outlines the process for transporting the *Enterprise* from Washington Dulles International Airport to the deck of the USS *Intrepid* at the Intrepid Museum, with a pit stop at John F. Kennedy (JFK) International Airport. Upon completion of the Intrepid Museum's new structure to house the *Enterprise* (which is currently several years away), the Orbiter must be moved again. The taxpayers have a right to know if NASA has determined that moving the *Enterprise* a total of three times will do any significant and irreversible structural damage to the Orbiter.

As you are aware, Congress recently passed an appropriations bill to fund NASA that also included a provision directing NASA to provide quarterly updates to Congress on the progress of the shuttle disposition. We are particularly focused on NASA's outlook on the Intrepid Museum's fundraising and construction efforts for the new display structure they plan to build, as well as details for the actual transport of the *Enterprise* to the Intrepid Museum.

More from Texas delegation

Robbie Kneviel makes the jump on the deck of the Intrepid. 2006

Various articles about Intrepid :

New York Barge Driver Smashes Wing of Enterprise
By Keith Cowing on June 4, 2012 3:35 PM. 40 Comments
Space shuttle sails through New York Harbor, CNN
"The space shuttle Enterprise took a journey more akin to
those of the aircraft carrier USS Enterprise than its orbital
sister ships on Sunday. The prototype shuttle floated on a
barge through New York Harbor, from John F. Kennedy
Airport en route to Bayonne, New Jersey."
Keith's note: What no one seems to be mentioning is that the guys driving/pulling the barge carrying Enterprise managed to smash the right wing of Enterprise into a dock. News
Tags: enterprise, space shuttle

A Concorde Is Disfigured While Parked in Brooklyn
Robert Stolarik for The New York Times

An accident last week sheared off the nose cone of the Concorde at Floyd Bennett Field.
By PATRICK McGEEHAN
Published: July 7, 2008

The supersonic passenger jet known as Alpha Delta retired unscathed in 2003 after nearly 30 years of speeding back and forth over the Atlantic Ocean. But in less than two years in Brooklyn, it already has had its pointy nose knocked off. In a multicultural crash in the middle of the night, the jet, a Concorde that is owned by a British airline, was hit by a truck that was hauling equipment from a Jamaican music and soccer festival. The truck clipped the distinctive nose cone off the parked Anglo-French jet about 3 a.m. last Monday, prompting an impassioned uproar among the jet's band of enthusiasts.

To many admirers, the tapered nose, which could be lowered up to 12.5 degrees to clear the pilots' field of vision during the jet's steeply angled takeoffs and landings, was what made the Concorde the Concorde. Within 20 hours of the accident, photos of the damaged plane appeared on the Internet, and Concorde lovers were deploring the level of care it had received during its postretirement odyssey in New York.

Bill White, the man responsible for maintenance of the 32-year-old jet, said he learned of the damage indirectly through the Concorde's fan club. The plane is in Mr. White's charge because he is the president of the foundation that operates the Intrepid Sea, Air & Space Museum. After British Airways retired its fleet of Concordes, the airline leased the jet to the museum, which displayed it on a barge in the Hudson River.

But when the Intrepid, the aircraft carrier that houses the museum, had to temporarily abandon its rotting pier on the West Side of Manhattan in late 2006, the Concorde had to go, too. British Airways considers the plane a three-dimensional billboard for its service, and it did not want the Concorde to stay with the Intrepid, which is docked at Staten Island and is closed while the West Side pier is being refurbished. So Mr. White found a temporary home for the jet at Aviator Sports, a recreational complex at Floyd Bennett Field in Brooklyn. The Concorde, which had traveled by barge to Manhattan from Kennedy International Airport in 2004, retraced most of that voyage en route to Aviator Sports. The jet's stay there has been more turbulent than a trans-Atlantic crossing at Mach 2. Shortly after agreeing to rent the plane for $15,000 a month during the Intrepid's absence, the original operators of Aviator Sports sold out. Their successors have balked at honoring the previous managers' pledge to pay for the plane's return to the Hudson waterfront, Mr. White said. The cost of that trip has been estimated at $250,000.

Mr. White said he was perturbed that the current operators of the complex did not notify him of the damage right away. He said he expected the cost of replacing the nose cone to be
covered by Aviator Sport's insurance policy.
He said he had been told that a driver "smacked the front end of the Concorde with his truck." He added that Intrepid museum officials would "take steps with Aviator to secure the aircraft so that this kind of a ridiculous happening won't happen again." Salvatore Musumeci, the director of security at the complex, said, "It was an unfortunate thing that happened, and everyone here is upset about it." He said he did not know who would pay for the repairs. John Lampl, a spokesman for British Airways, said the airline expected the jet to be fixed and returned to the Manhattan pier by the end of the year. Replacement parts for Concordes are no longer being manufactured, but Mr. Lampl said the airline knew of collectors who had bought spare nose cones at an auction. He said he expected that Intrepid officials could buy one from one of those enthusiasts.

Dead Boxer's Gift Of Life
BY MARTIN MBUGUA DAILY NEWS STAFF WRITER WITH
J.K. DINEEN
Wednesday, July 04, 2001
Boxer Beethavean (Sweet Bee) Scottland, comatose for six days after he was knocked out in a fight aboard the Intrepid Sea-Air-Space Museum, died of his injuries Monday and his organs were removed for donation. Scottland's widow, Denise, who kept vigil at his bedside, drove back to their Maryland home and three children, ages 2, 5 and 7, yesterday. Scottland, 26, of Landover Hills, Md., hit the canvas in the 10th round of a nationally televised fight with

light heavyweight George (Khalid) Jones of Paterson, N.J., aboard the Intrepid on June 26.
He was taken to Bellevue Hospital, where he twice underwent surgery, remaining comatose until his death at 10:36 p.m. Monday. His wife, mother, mother-in-law and several brothers and sisters were at his bedside.

Space shuttle Enterprise suffers damage in ferocious storm Prototype now sits exposed as inflatable pavilion, lashed by winds, collapses in NYC Below:

Denise Chow / Space.com
The space shuttle Enterprise is seen after Hurricane Sandy at the Intrepid Sea, Air and Space Museum in New York on Tuesday. Photo shows the shuttle's protective shelter has collapsed and the orbiter has incurred some damage.
By Robert Z. Pearlman

updated 10/30/2012 4:14:08 PM ET

Space shuttle Enterprise, NASA's original prototype orbiter, is sitting exposed and appears to have been partially damaged by Hurricane Sandy after the severe storm passed over New York City on Monday night.

On public display aboard the Intrepid Sea, Air and Space Museum, a converted aircraft carrier, since July, Enterprise had been protected from the elements inside a pressurized pavilion. Based on photos posted online, the inflatable structure appears to have first deflated and then been torn by the winds of the now post-tropical storm cyclone.

Photographs show the 180-foot-long (55 meters) by 60-foot-high (18 meters) pavilion's cloth exterior now lies draped over Enterprise, although much of the shuttle's nose section and part of its payload bay is uncovered. The orbiter's vertical stabilizer, or tail, is protruding out of the top of the fabric, where it appears part of the spacecraft has been torn away.

Intrepid officials did not immediately respond to a request for comment.
Space news from NBCNews.com

NOAA / NASA GOES Project

Storm damages space shuttle Enterprise

The "superstorm" Sandy flooded Pier 86, where the Intrepid is anchored, submerging part of the museum's main entrance under water. Similar extensive damage was seen throughout the city and region, leaving buildings destroyed, millions of people without power and at scores dead. [Superstorm from Space: Hurricane Sandy Satellite Photos]
Enterprise was delivered to the Intrepid in June after being transferred from its previous home at the Smithsonian National Air and Space Museum's Steven F. Udvar-Hazy Center in northern Virginia. In its place, the retired space shuttle Discovery is now on display at the Udvar-Hazy Center, having arrived in April after NASA retired its 30-year shuttle program in 2011.
Discovery was also in Sandy's path, and the Smithsonian remains closed due to the storm. However, no damage to that orbiter was reported, nor was any damage evident on webcam footage of the shuttle.

Enterprise, built in the 1970s, never made it to space, but it was used as a prototype to test the space shuttle design during approach-and-landing glide tests.
Since its arrival at the Intrepid, Enterprise has been housed in its climate-controlled steel-and-fabric "Space Shuttle Pavilion." This shelter was never meant to be permanent, however. Eventually, Intrepid plans to build a larger facility to showcase the shuttle and enhance its other space exhibits and educational displays.

 As of Tuesday morning, superstorm Sandy was centered about 120 miles (190 kilometers) east-southeast of Pittsburgh, slowly moving westward and weakening over Pennsylvania, according to the National Hurricane Center (NHC) in Miami.

Thank you for taking the time to read "NASA and the Shuttle Shuffle. The author will entertain any questions that you may have at:
Valkyr1097@nyc.rr.com

www.ingramcontent.com/pod-product-compliance
Lightning Source LLC
Chambersburg PA
CBHW041102180526
45172CB00001B/68